The Ohio & Erie Canal
A Glossary of Terms

The Ohio & Erie Canal

A Glossary of Terms

Second Edition, Revised and Expanded

ॐ

Compiled by Terry K. Woods

The Kent State University Press
KENT, OHIO

© 2011 by The Kent State University Press, Kent, Ohio 44242

Library of Congress Catalog Card Number 2011016819
ISBN 978-1-60635-106-2
Manufactured in the United States of America

Library of Congress Cataloging-in-Publication Data

The Ohio & Erie Canal : a glossary of terms / compiled by Terry K. Woods. —
2nd ed., rev. and expanded.
p. cm.
ISBN 978-1-60635-106-2 (pbk. : alk. paper) ∞
1. Ohio and Erie Canal (Ohio)—Terminology.
I. Woods, Terry K., 1937–
II. Title: Ohio and Erie Canal.
TC625.O28W66 2011
627′.1309771—dc23
2011016819

British Library Cataloging-in-Publication data are available.
15 14 13 12 11   5 4 3 2 1

Contents

ॐ

Introduction to the First Edition

୬ଡ

A casual reading of current newspaper and magazine articles concerning operating days on the Ohio & Erie Canal reveals both a number of duplicate terms meaning different things and different terms meaning the same thing. Deeper research into official reports and early newspaper accounts as well as interviews with exboatmen only compounds the problem. Engineers, it seems, called things by one name, boatmen called them by another (or two or three), and the casual citizen something else.

This booklet is an attempt to arrive at the most commonly used terms for structures, artifacts, and other canal-related items. I have assumed that most people delving into the history and lore of Ohio's canal era will eventually get around to reading the official reports of the state canal commissioners and their successors, the Board of Public Works, and I have tried to use common terms that appear most often in those documents. But here, too, there are multiple usages of some terms and multiple terms for some items. What something was called seems often to depend upon when, where, and by whom the reports were written.

Most of the common terms in this glossary have come from the 1909 Board of Public Works Report and accompanying map of the Ohio & Erie Canal. This document, issued near the end of the attempted rebuild of the canal, mentions nearly every structure and artifact of the canal as it then existed. Since operation on the canal was practically over by that time, further changes in terminology would thus come from people not intimately involved with day-to-day operations.

This glossary lists a primary term for each item. Secondary terms for the same item are also listed, along with the often more colorful terms applied by the boatmen. There are undoubtedly going to be favorite terms of some researchers and students that have been given the "wrong" meaning here. And to others, the meanings will be horribly garbled. Please let me know all of your comments. Through them, we can get closer to a truly accurate glossary of terms used along the Ohio & Erie Canal.

Terry K. Woods

Introduction to the Second Edition

༰ঞ

It has been interesting (and fun) redoing this book and seeing what an additional fifteen years of research can add to this glossary. We have not changed any of the primary definitions that appeared in the first edition, but, as in the case of lock keepers in Ohio and the name of mule drivers in general, we have added quite a bit of historical information. And we have fine-tuned much of the historical information in the original definitions.

There was good feedback from readers of the first edition, and we have made the necessary corrections in this one. Plus, we have added a few definitions, such as "transom" and "widewater," that somehow eluded us the first time around.

We have also added a few contemporary canal photos to this edition and included a comprehensive description of the route of the Ohio & Erie Canal from Cleveland to Portsmouth. By detailing the water sources, sidecuts, and branches of the longest of Ohio's canals, we hope to bring this famous waterway into sharper focus for a new generation of Ohio & Erie Canal–era buffs.

I urge those who read this new edition, like those who read the first, to tell us where we "hit the mark" with our definitions and history and where we need to improve. Together we can achieve a truly accurate glossary of terms used along the Ohio & Erie Canal.

The Canal Route

When the entire length of the Ohio & Erie Canal was opened for navigation in October 1832, it was 308.14 miles long (exclusive of feeders) and employed 146 locks to overcome a total difference in elevation of 1,207.35 feet between Lake Erie, the two summits, and the Ohio River. It also contained five guard locks and 153 culverts. The canal crossed streams fourteen times upon aqueducts and eight times in slackwater pools above dams. Six of these crossings were also used as water supplies.

Two sloop locks, each of six-feet lift with chambers of 25 feet by 100 feet, joined the canal to the Cuyahoga River about a half mile above its junction with Lake Erie. These locks allowed the largest class of sloops then operating on the lake to enter the canal and Merwin's Basin to pick up or off-load cargo at the many warehouses lining the shore.

A very short, privately financed canal, dubbed the Ohio City Canal, was constructed inland from the west bank of the Cuyahoga River, giving Ohio City's industry access to the Ohio & Erie Canal via the Cuyahoga River later in the 1830s.

From here, the Ohio & Erie Canal initially used 42 locks to overcome a difference in elevation of 389½ feet along the Cuyahoga Valley to the Portage Summit at Akron. There were three aqueducts in this stretch of canal. One crossed Mill Creek nine miles from Cleveland and another crossed Tinker's Creek thirteen miles from Cleveland. The third, at Peninsula, carried the canal from the left to right bank of the Cuyahoga River. During 1837, Lock No. 42, the original Four Mile Lock, was removed and Five Mile Lock, No. 41, reduced to a four-foot lift. During the late 1870s, when the northernmost three miles of the Ohio & Erie Canal were given to the City of Cleveland, a new outlet lock was constructed near the foot of Dille Street and became the "new" Four Mile Lock, No. 42.

South of Lock No. 23, the canal abandoned the valley of the main Cuyahoga River for that of the Little Cuyahoga, which it followed to the vicinity of Lock No. 16. It then followed the old northern outlet of the summit pond (now Summit Lake) through the new town of Akron to the beginning of the Portage Summit level, some 38 miles from Cleveland. The privately financed Pennsylvania & Ohio Canal entered the Ohio & Erie Canal here through an ascending lock into the lower basin adjacent to Lock No. 1.

The canal received four feeders between Cleveland and Akron. The Pinery Feeder just below Brecksville and another just below Lock No. 31, about a half mile below the Peninsula Aqueduct, were both short feeder channels from dams across the main Cuyahoga River. Two feeders were also taken from dams across the Little Cuyahoga: one just below Lock No. 21 and another just below Lock No. 16. Furnace Run, above Lock No. 27, was initially crossed on slackwater. This crossing was later converted to a one-span iron aqueduct.

The Portage Summit level was fourteen miles long and passed along the eastern edge and through the southern tip of the summit pond. Canal engineers were forced to erect a floating wooden bridge across that part of the pond for the towing animals when the deep muck and quicksand in the area made it impossible to build up an embankment or to sink pilings into the pond's bottom. A system of natural lakes, increased in capacity by dams over the years (the Portage Lakes), fed the summit.

The first lock on the Tuscarawas slope was located on the west (right) bank of the Tuscarawas River at the southern edge of the present-day city of Barberton. A short quarter of a mile south of Wolf Creek Lock (No. 1) the canal crossed Wolf Creek on slackwater behind a dam (later converted to an aqueduct). The canal remained on the right riverbank to just above Clinton where it crossed to the left bank upon a dam's embankment. The canal stayed on the left bank of the Tuscarawas for the next 28 miles, through the towns of Fulton and Massillon. Initially, there was just one lock in Massillon—Lock No. 5 just south of town. It was originally built with a twelve-foot lift. During 1838 and 1839, the lock was reduced to a six-foot lift, the mile of canal just north of that lock was straightened, and an additional lock with a six-foot lift constructed just south of what would later become Walnut Street. Then, during the early 1900s canal rebuild, the original lock was removed, the canal bank raised, and a new, concrete lock built about a mile below.

The canal continued on the left bank of the Tuscarawas through Rochester, Navarre, and Bethlehem (combined into the village of Navarre in 1876) before again crossing the river, this time upon a three-span aqueduct, a short distance above Bolivar. The privately financed Sandy & Beaver Canal connected the Ohio & Erie Canal at Bolivar with the Ohio River at Glasgow, Pennsylvania. Later, the westernmost six miles of this canal were acquired by the state as a feeder to the Ohio & Erie Canal.

A short canal to an iron foundry was constructed by the Zoarites (members of the Separatist Society of Zoar). It entered the Ohio & Erie Canal from the west near Lock No. 7 below Bolivar. The Zoar Sidecut enabled boats to enter the village of Zoar once they had been poled from the Zoar Feeder (just below Lock No. 10) and across the river.

The New Philadelphia Lateral exited the Ohio & Erie Canal in a slackwater pool behind the Sugar Creek Dam below Dover. The three-and-a-half-mile-long navigable Trenton Feeder entered the Ohio & Erie Canal just below the village of Trenton (Tuscarawas).

The canal continued along the right bank of the river for 57 miles from Bolivar to Roscoe (now part of Coshocton) where the Walhonding River joins the Tuscarawas to form the main Muskingum. On its way, the canal passed through the towns of Dover (later Canal Dover), Lockport (now part of New Philadelphia), New Castle, Trenton (Tuscarawas), Gnadenhutten, Seventeen, Port Washington, Evansburg (Orange), and (Canal) Lewisville. Sugar Creek, just below Dover, was crossed upon a slackwater dam. The Walhonding River was crossed upon a five-span aqueduct just above Roscoe. Originally, this aqueduct entered the lower of three basins (Roscoe Basin) through a fourteen-foot lift lock (No. 26). Later, the lock was rebuilt into a double lock with two chambers, each having a lift of seven feet. This resulted in some confusion of the lock numbering system below, as the state engineers considered it one lock, while many references call it two—Nos. 26 and 27. For the next 14 miles below Roscoe, the canal followed the right bank of the Muskingum River to the canal's lowest point between the Portage and Licking Summits. There were originally 29 locks between the Portage Summit and this point to negotiate a 238-foot fall in elevation. Water was obtained in this section from Summit Lake, the slackwater crossings of Wolf Creek, and the Tuscarawas River above Clinton. Water was also obtained from two small streams between Clinton and Massillon. After 1847, feed water was supplied to the canal at Massillon from Sippo Creek. There was a short feeder from the Tuscarawas River opposite Zoar and a slackwater crossing of Sugar Creek just below Dover. The Zoar feeder was so dependable that no water was required at the Sugar Creek crossing. A three-mile-long navigable feeder entered the main canal just south of Lock No. 16 below Trenton and was known as the Trenton Feeder. There was also a mile-long feeder from the Walhonding River into the Lower Basin above Roscoe. In the early 1840s, the 25-mile-long Walhonding Canal was built up the valley of the Walhonding River from Roscoe.

Before leaving this valley, at the long-gone town of Webbsport, a three-mile-long canal, the Dresden Sidecut, was built in 1831 south to the village of Dresden to connect the Ohio & Erie Canal to the Muskingum River trade.

The Ohio & Erie Canal now left the valley of the Muskingum and ascended the valley of Wakatomaka Creek for about nine miles before passing through a gap in the divide and ascending the valley of the Licking to the summit in Licking County, encountering the towns of Hartford, Frazeyburg, Nashport, Lickingtown, and Newark. The total length of this section was 32 miles and had an ascent of 160 feet, which required 19 locks to overcome. There were three aqueducts in this relatively short stretch. One was a single-span affair across the Wakatomaka Creek about a mile above Frazeysburg; two were in Newark, one across the North Fork of the Licking and one across Raccoon Creek.

One of the more unusual features of the Ohio & Erie Canal was southeast of Newark at the "Narrows of the Licking." Here, the Licking River was forced to flow through a narrow, deep, and rocky defile leaving little room along the riverbanks to construct a canal channel. The engineers solved this problem by damming the river at the eastern end of the gorge and leading the canal through the valley in the resulting slackwater pool. At one point, it was necessary to blast a towpath along the face of a sheer rock wall, obliterating a pictograph (attributed to early aborigines) of a large black hand that gives the area its name to this day. This slackwater pool was one water source for this section of canal; two others were a feeder from the north fork of the Licking River at Newark and a feeder from Raccoon Creek.

The Licking Summit was 14 miles long and touched the towns of Heath (where the National Road crossed the canal), Hebron, and Millersport. Three aqueducts were located on the summit level. One aqueduct was located at Hebron, another two miles south of Hebron, and the third across the south fork of the Licking, some three miles below Hebron.

Water was supplied to the summit level by the navigable "Granville Feeder" near the northern end of the level and by the Licking Reservoir, now known as Buckeye Lake. The Licking Reservoir initially occupied a natural basin nearly eight miles long by about a half mile wide and covered some 2,500 acres. An artificial embankment nearly four miles long was constructed to impound the water and two miles of that embankment formed the canal's towpath. A six-mile-long feeder, taken from the south fork of the Licking River at Kirkersville, passed over the canal on an aqueduct and then entered the reservoir at the southwestern end to augment the natural drainage into the basin.

A 500-acre addition called the New Reservoir was constructed in 1836. It was located on the west side of the original towing path embankment. There were apparently some difficulties in obtaining an even flow of water from the old to the new reservoir. During the lease of the canals in the 1860s, Archimedes-type screw pumps, 24 inches in diameter, were employed to raise water from the old to the new reservoir during periods of low water. Two guard locks, one at each end of the summit level, were constructed when the reservoir was enlarged.

In order to have the summit level low enough to get the maximum benefit from the reservoir, it was necessary to dig a three-mile-long deep cut through the dividing ridge. The cut began near the Kirkersville Feeder Aqueduct, extended to the south, and was 34 feet deep near the center.

From the Licking Summit, the canal descended southward along the valley of Walnut Creek for some 18 miles, passing through the towns of Baltimore, Havensport, Carroll, Lockville, Waterloo, Canal Winchester, Raineysport, and Sharp's Landing. The canal crossed and recrossed Walnut Creek, taking advantage of the terrain to maintain a level, before entering a plain between the valleys of Walnut and Big Belly Creeks. The canal then followed the Big Belly to Lockbourne where that creek intersected with the Scioto River. This stretch of canal was some 30 miles long and used 30 locks to descend 202 feet in elevation.

A "wet weather" feeder from Walnut Creek entered the canal about seven miles from the summit. A feeder from Blacklick and Ragers Run was introduced some twelve miles below the summit. Water was also introduced into the canal at Carroll by the Lancaster Lateral Canal, a private branch canal that was later taken over by the state, extended, and renamed the Hocking Canal. The Columbus Feeder Canal entered the Ohio & Erie Canal at Lockbourne.

From this point to the Ohio River near Portsmouth, the canal generally followed a southerly course through the Scioto Valley, touching the towns of Holmes' Landing, Millport, Bloomfield, Circleville, Westfall, Yellowbud, Deer Creek, Andersonville, Clinton Mills, Chillicothe, Sharonville, Waverly, Jasper, Cutler's Station, and, by crossing the river after leaving the canal, Portsmouth. The first 15 miles of this section ran along the east side of the Scioto, crossing Walnut Creek near Bloomfield, about seven miles below Lockbourne, in slackwater above a dam. At Circleville the Scioto River was crossed upon a seven span aqueduct. A lock with a nine-and-a-half-foot drop was built into the stonework of the aqueduct's western abutment.

Eleven miles below Circleville, the canal crossed Yellow Bud Creek upon an aqueduct and, 14 miles below Circleville, crossed Deer Creek upon another aqueduct. Another aqueduct was used to cross Paint

Creek, some two miles below Chillicothe. Fish Creek, 21 miles north of the Ohio River, was crossed on a twin-arch culvert. Two more stone culverts were in this last stretch of canal—a one-arch culvert across Camp Creek and a three-arch culvert across Scioto Brush Creek eight miles north of the canal's junction with the Ohio River. The total length of this section was 87 miles. Twenty-four locks were required to lower the canal 211 feet. An outlet lock allowed canal boats to venture out onto the Ohio River. This portion of the canal received water from three sources: from the slackwater crossing of Walnut Creek, by a feeder dam on the Scioto about two miles below Circleville, and from another feeder dam on the Scioto some four miles further down the river. In addition, a navigable hydraulic canal intersected the canal at Chillicothe.

Originally, the Ohio & Erie Canal terminated across the Scioto River from Portsmouth on the right, or western, bank. There was to have been a channel through the isthmus separating the Scioto from the Ohio River and a lock into the Scioto River itself, to allow canal boats to cross and enter Portsmouth. When this sidecut was begun, however, the Scioto River poured through it, changing its junction with the Ohio River and leaving the canal on the western bank of the Scioto. Numerous attempts to bridge the Scioto or make a branch canal on the east bank resulted in failure. Much later, in 1887, a double lock was constructed into the Ohio River from the old Scioto channel, but traffic on that section of the Ohio & Erie Canal amounted to very little by then and the lock was seldom used.

Terry K. Woods
6939 Eastham Cir NW
Canton, Ohio 44708-1008

The Ohio & Erie Canal
A Glossary of Terms

ॐ

Glossary

❧

ABREAST HITCHING: (See TANDEM HITCHING.) Hitching the towing animals of a canal boat side by side. This method of hitching was used on a number of eastern canal systems that had wide towing paths. It was not common on the Ohio & Erie Canal, as the ten-foot wide towpath and heavy traffic made it nearly impossible for two teams to pass. More power could be generated using abreast hitching, however, so some boatmen resorted to using it during the later operating days of the canal when traffic was light.

ABUTMENT: End support on which a bridge, dam, or aqueduct rested.

ALLIGATOR LOCK: (See LOCK.) A type of lock that had earthen embankments, protected by slope wall and riprap, for the chamber. Only those portions of the lock that contained gates were constructed in a normal manner of stone, concrete, or wood. Many guard locks along the Ohio & Erie Canal were of this type of construction.

ANIMALS: The propelling power, for the most part, of canal boats on the Ohio & Erie. Though both horses and mules were used, mules were often preferred as they were perceived to be smarter and longer lived. A horse could only stand a few years of towing canal boats. Boatmen were also addict horse traders, seldom keeping the same set of spare animals an entire season.

Walhonding Aqueduct near Coshocton. (From the author's collection.)

AQUEDUCT: A structure for carrying the canal channel and towing path across a stream or valley too wide or deep for a culvert. During the initial building of the Ohio & Erie, a wooden trough with a towing path constructed along one side carried the waters of the canal from one side of the stream or valley to the other. Usually this trough was anchored to the stream's bank with stone abutments. Spans longer than thirty feet were supported by stone piers. Some later updated aqueducts employed an iron trough, and the early 1900s rebuild of an aqueduct across Wolf Creek near present-day Barberton was concrete. There was an all-wooden aqueduct on the branch Sandy & Beaver Canal at Bolivar.

BALANCE BEAM: A long, wooden beam projecting out as part of the top of a miter gate that assisted in balancing the weight of the gate so that it was easier to operate. The gates were opened and closed by manual operation of the balance beams. The term *sweep* was also used, though less frequently.

The balance beams on the lock's miter gates can be easily seen in this photo
of a three-cabin freighter as it locks through Paper Mill Lock in Massillon.
(Photo from the collection of the Massillon Museum.)

BAR: An obstruction in the canal's channel, generally caused by streams
that flowed into the canal, bringing in silt and sand. Boatmen named
bars that occurred frequently at the same location. There was Buttermilk
Falls Bar above Seventeen Mile Lock, Granny Run Bar above Deep Lock,
and Cemetery Run Bar at the Massillon Cemetery.

BARGE: A nonsteerable craft pulled or pushed by a boat. On the Ohio &
Erie Canal, barges were infrequently used by crews of state boats for
conveying materials. Barges were not used for regular transport of cargo
or passengers on this canal.

BASIN: A cutout or enlargement of the canal channel, usually along
the berm bank, that gave boats access to canal-side industries. A basin
could also be located along a towpath bank that employed a removable
towpath bridge.

BATTENED: To protect a surface by the addition of vertical or horizontal wooden strips.

BATTER: A slight slope, from top to bottom, of a stone wall, such as the interior of a lock, to counteract outward pressures exerted by alternate freezing and thawing of the earth.

BERM or **BERM(E) BANK:** The engineering or technical term for the bank of the canal channel opposite the towpath. This is the name most often seen in official written reports. Boatmen, however, nearly always referred to this bank as the heelpath because it was opposite the towpath. When the canal was first built, the berm bank was often allowed to follow the natural contour of the land when it would reduce the cost of construction. This resulted in the canal's width extending up to 150 feet or more on occasion.

BOARDING BOAT: Refers to a craft, usually a converted freighter, that provided housing for the work crew of a State Dredge during the early 1900s rebuild.

BOAT: (See CANAL BOAT.) Also used as a verb, "to boat," on the canal.

BOATMAN: Each man (and possibly woman) who owned or worked on canal boats. The term *canawler* (in all its various spellings) was seldom used.

BOATYARD: A place along the canal, often in conjunction with a drydock, where canal boats were constructed. Although canal boats were built at many locations along the Ohio & Erie Canal and its branches—as far as away as Urichsville on the Big Stillwater—most were constructed at Peninsula and Boston.

BOW: The forward portion of a canal boat.

BOWSMAN: A crew member on a canal boat whose duties included leaping off the craft at a lock to ease the bowline around a snubbing post, which arrested the craft's forward momentum. This kept the boat from crashing into the far gates and centered the craft within the lock chamber while it rose or fell in elevation.

BRANCH CANAL: (See LATERAL CANAL, SIDECUT, and SLIP.) A side channel that extended from the main canal and connected with a town or industry that the main canal did not reach.

BRIDGE PLANK: A cleated plank carried on board a canal boat that was employed to walk a team from the stable cabin to the towpath and back.

CABIN: The enclosed space(s) on a canal boat. In boat-building jargon, a cabin was referred to as the "house."

CANAL BOAT: (See FREIGHTER, LINE BOAT, PACKET, SCOW, and STATE BOAT.) A steerable craft, usually of frame construction, generally towed by animals. Ohio & Erie canal boats developed from the Erie's bull-head design and were about 80 feet long by 14 to 14½ feet wide. Up until the mid-1840s, boats were sharper-prowed, easier to tow, had a loaded draft of approximately 3 feet, and carried around 50 tons. Later designs had the boats "fleshed out" with loaded drafts up to 4 feet and cargo capacities of 80 tons or more.

CANAL GRASS: An underwater growth of weeds. One of the big maintenance headaches of the state boat crews was the need to continually cut this aquatic growth. If left unchecked, it would obstruct navigation and even restrict the flow of water for hydraulic purposes. This grass often had to be cut four or five times per season.

CANAWLER: (See BOATMAN.)

CAPTAIN: The boss of a canal boat. In the early days of boating on the Ohio & Erie, to be a captain of a line boat was an honor, indeed, and a position of some standing in the community. During the later operating days, however, the title had degenerated to be primarily an honorary one that usually referred to the owner/operator of a canal boat.

CATWALK: A narrow walkway connecting the decks of all three cabins of a three-cabin freighter. A boatman could walk from one end of the boat to the other using the catwalk and decks when the midships were filled with cargo.

CHAIN: A surveyor's measurement often used when laying out the original canal. One chain = 66 feet.

CHANGE BRIDGE: A bridge carrying the towpath from one side of the canal to the other when the towpath changed sides due to a river crossing.

CLEARING: The act of removing all brush, saplings, heavy timber, etc., more than one foot high for a distance of twenty feet on each side of the grubbed area for canal construction.

COMPOSITE LOCK: A lock consisting of a rough stone structure lined with plank to make it watertight. There is no record of this type of lock being used on the Ohio & Erie Canal. Some use of this design was made on the Sandy & Beaver and Pennsylvania & Ohio, two privately built connecting canals.

COOK: The member of a canal boat crew who obviously did the cooking. Usually a female, during the later operating days the cook was often the wife of the captain. Sometimes the title *cook* was a euphemism for the captain's female companion.

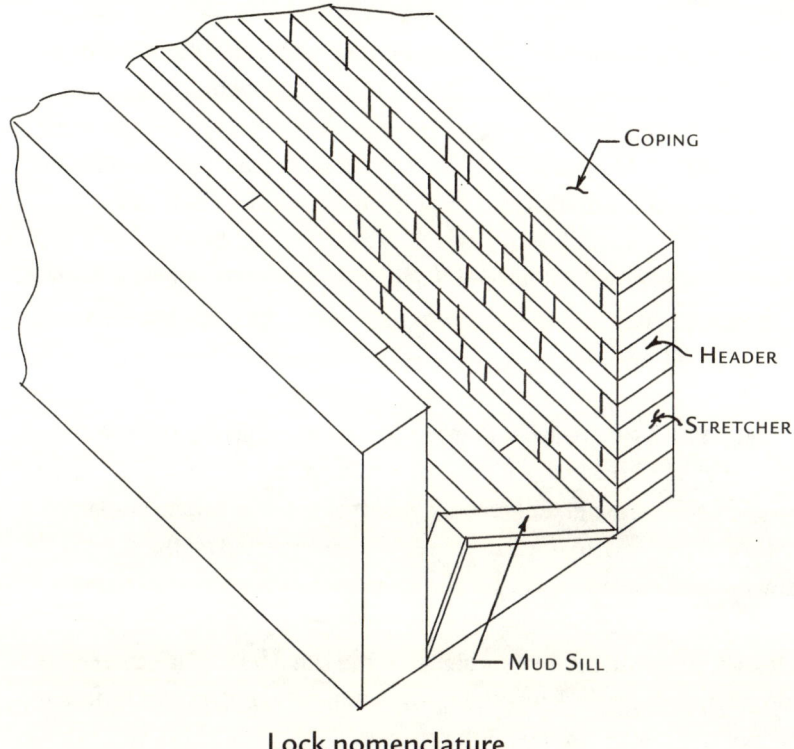

Lock nomenclature

COPING: The capstone, or stones on the top of a stone lock. Early specifications called for these stones to be "not less than three feet in width, well cut, jointed and bedded."

CREW: The people regularly employed on a canal boat. During the early operating days on the Ohio & Erie Canal, a standard freight boat crew consisted of a captain, two steersmen, a bowsman, two drivers, and a cook. A packet's crew consisted of a captain, two steersmen, two bowsmen, two drivers, a cook, and a steward. Packet boats ran day and night; freighters often did not. Toward the end of the canal's life, when passenger packets had long disappeared, a minimum crew for a freight boat was two or three people: a steersman and a driver, and possibly a "hired man."

CRIB: A boxlike wooden framework filled with rock. Rock-filled cribs were often used to support dams and other engineering structures.

CULVERT: A barrel-shaped structure constructed of stone or wood that allowed a narrow stream or gully to pass under the canal.

DAM: A wall-like structure built across a stream from stream-bottom to a prescribed height that impounded water behind it (in a slackwater pool) to that height.

DEADEYE: A cast-iron bar or eyelet mounted at the edge of the top deck of a canal boat to which the towline was attached. There was a deadeye on each side of the boat, located a bit aft (depending upon the owner's theory of physics), to minimize the side force exerted on the craft by the towline and make it easier for the steersman to maintain a course.

DECK: The flooring, or walking area, of a canal boat. On a three-cabin freighter, the bow deck was the roof of the bow cabin, the stable deck the roof of the stable cabin, and the stern deck the roof of the stern cabin. The roof of a line boat or packet was also considered a deck. The flooring inside a line boat or packet or in the holds of a three-cabin freighter does not appear to have been called a "deck."

DOUBLE LOCK: Two lock chambers built in the same structure, one after the other. On most other canal systems, a double lock was two chambers built side by side in the same structure. The one double lock on the Ohio & Erie Canal was Lock No. 26 near Roscoe, which was originally one chamber with a fourteen-foot lift located adjacent to the Walhonding Aqueduct. It was changed to two chambers, each with a seven-foot lift, when it was discovered that emptying the fourteen-foot-lift lock chamber caused the aqueduct to go dry occasionally.

One of five steam dredges built by the state during refurbishing of the Ohio & Erie Canal in the early 1900s. (From the author's collection.)

DOUBLING: Being able to navigate a lock in its found condition. If the last boat through the lock was traveling in the opposite direction, your boat could enter that lock without having to change the positions of the near gates or paddles. That lock could be doubled.

DRAFT: The depth to which a boat's hull would be immersed in the water of the canal.

DREDGE: A steam shovel mounted on a compartmentalized barge that could be broken down into smaller sections and taken through locks and under bridges. Initially, Ohio's canals were drained each spring while teams of workers dug out bars and refurbished the channel for the coming boating season. When the canals were leased to a consortium in 1861,

Boy driver and stopped team near Lock No. 17 north of Akron. (From the author's collection.)

the lessees built steam dredges and were able to keep the channels clear without draining and stopping traffic for long periods. When the state regained maintenance responsibility for the canals in 1878, the dredges were obtained from the lessees and the practice of dredging continued. The state built its first dredge and operated it on the Massillon canal section in 1909.

DRIVER: The person who would drive the animals along the towpath. During the later operating days of the canal, this would often be the young son or daughter of the boat's owner/captain.

DROP GATES: Upper lock gates that pivoted (dropped) down onto the lock chamber floor when the head of water in the chamber was reduced sufficiently to allow it. The gate was then winched back into an upright position and fastened into place so that the next boat could lock through. This type of gate may have been used to some extent on the privately financed Sandy & Beaver Canal but not on the Ohio & Erie.

DRYDOCK: A cutout adjacent to the canal bank, usually on the river side of the canal, plank-lined, and just wide enough to hold two boats side by side. Drydocks were used primarily to repair and recaulk older boats. The drydock would be filled with water from the canal, and the boat floated in through a bridged gate in the towpath. The boat would then be positioned over a set of wooden sawbucks on the bottom of the drydock and the water allowed to drain through a gate at the far side of the drydock into a lower level of the canal or a nearby watercourse. The boat would then settle onto the bucks as the water drained out. The drydock would again be flooded with water from the canal to get a repaired boat back into service.

EMBANKMENT: The method employed to construct a canal channel through an area with an initial elevation below that of the proposed channel bottom.

EMINENT DOMAIN: This provision, embedded in the 1825 act to build Ohio's canals, gave the canal commissioners the ability to "occupy" a parcel of land and, by constructing a canal on it, to obtain a Fee Simple Title to that parcel.

EXCAVATION: The method employed to construct a canal channel through an area with an initial elevation above that of the proposed canal channel bottom.

FEEDER: A channel directed into the canal from a water source (reservoir, slackwater pool, etc.) for the express purpose of providing a required supply. Several feeders to the Ohio & Erie were navigable—notably the

Tuscarawas Feeder at Trenton (three miles long with one guard lock) and the Walhonding Feeder (one mile long with one guard lock).

FEEDER GATES: Structures built across the outlet of feeder channels, very much like sluices, except that they worked in reverse: they regulated the amount of water from feeders into canal channels. These structures have also been called "head gates" in some reports.

FIT A LOCK: The term used by boatmen for preparing a lock for the continued travel of a boat. When no formal locktender was on duty, a crew member (often a younger family member during the later operating days of the canal) would run ahead of the boat to ensure that it was fit—in other words, that the gates and chamber water level were correct for passage of the boat. Also called "make a lock." (See LOCKING THROUGH.)

FLIGHT OF LOCKS: A set of separate but closely spaced locks.

FLOODGATES: (See SLUICE.)

FOUNDATION: The support for masonry structures such as locks and culverts where no bedrock could be easily reached. One-foot-square oak timbers, placed an equal distance apart, were laid across the width of the structure extending beyond the walls of the lock or culvert. Three or more rows of sheet piling were then driven underneath the foundation. The spaces between the timbers were filled with well-worked puddle. The top of the timbers and puddle was then covered with three- or four-inch planking, well treenailed. The walls of the lock or culvert were then erected upon this platform.

FREIGHTER: A canal boat built to carry freight. The totally enclosed line boat or two-decker was a freight boat that could also carry a passenger, or a few, on occasion. During the later operating days on the

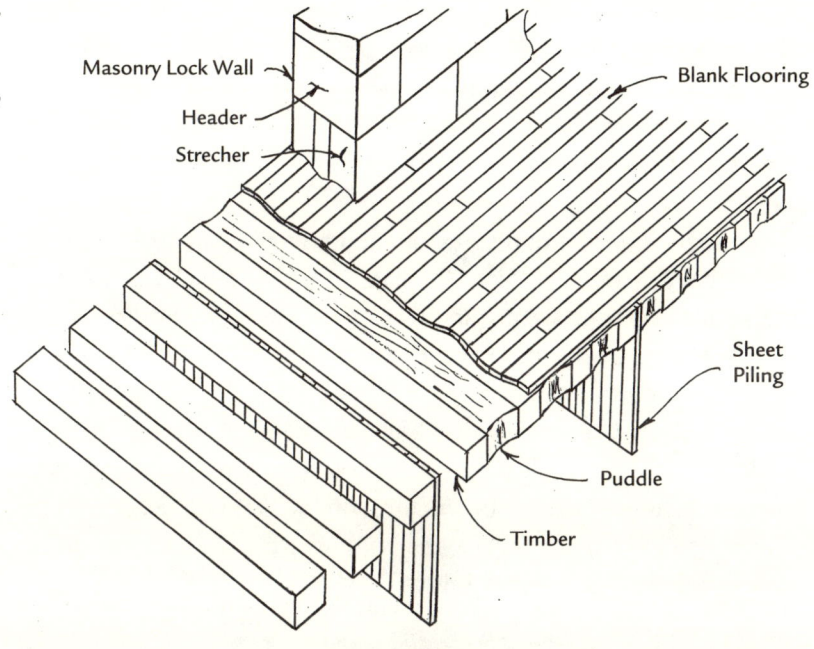

Masonry Lock Wall

Header

Strecher

Blank Flooring

Sheet
Piling

Puddle

Timber

Foundation

canal, the most common craft on the Ohio & Erie was the three-cabin freighter. After it had been proven uneconomical to maintain way stations along the canal's route for the exchange of towing animals, most boatmen began carrying their spare animals on board. As the formal freight lines disappeared after the early 1870s, most bulk cargo was carried by independent haulers, thus creating the need for three cabins. The stern cabin housed the captain and his family, the center stables for animals, and the bow cabin for any nonfamily crew members. Cargo was carried in the twin midships holds located between the three cabins.

GATE CHAMBER: A recess built into the walls of a lock into which open gates could fit. This gave a boat the entire fifteen-foot width of the lock chamber.

GOON NECK: (See GOOSENECK STRAP.)

GOOSENECK STRAP: The metal strap holding the heel post of a miter lock gate to the top (coping) of the lock.

GROUND CULVERTS: (See LOCK CULVERTS.)

GRUBBING: The act of digging out and removing all vegetation and roots from a distance of thirty-three feet from the canal's centerline on the towpath side and twenty-seven feet on the opposite side (for a total of sixty feet) of a projected canal line.

GUARD GATE: A single pair of gates at the head of a feeder into a navigable canal that could be closed to protect the canal from high water. There were at least two such structures on the Ohio Canal, one on the Pinery Feeder near Brecksville and one a converted guard lock on the Zoar Feeder in Tuscarawas County.

GUARD LOCK: A lock at the influx of a slackwater impoundment that did not provide a specific lift. Instead, the guard lock raised a boat to the momentary level of the slackwater impoundment and guarded the lower canal against high water. Guard locks were often constructed with an earth chamber protected by riprap, and only the gate support areas were constructed in the normal manner of stone, wood, or concrete.

GUNNEL: (See GUNWALE.)

GUNWALE: A narrow (approximately twelve inches wide) protrusion about the periphery of a canal boat. Boatmen could walk along the gunwale from one end of the boat to the other when cargo was in the midships.

HATCH: A trapdoorlike device in the deck that allowed access to the cabins of a canal boat by way of a wooden ladder, which was permanently attached to the cabin wall under the hatch.

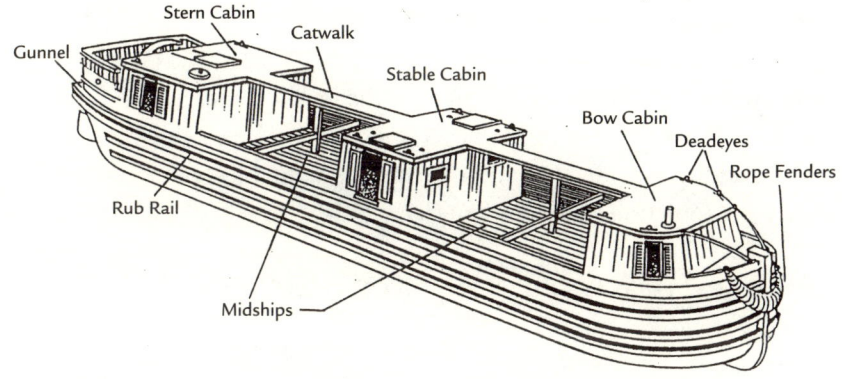

Three-cabin freighter

HEAD: Refers to a height of water in a channel or lock chamber. This term can be related to water pressure.

HEADER: A building stone, as in a wall of a lock chamber, laid so its long axis was perpendicular to the face of the wall. Its purpose was to tie the rough and finished stone walls together into one massive structure. Specifications set down in 1825 called for headers to be placed a minimum of ten feet apart in any stone course. Their minimum dimensions were to be 12" x 24" x 4½'. Later specifications called for headers to be a minimum length of five feet.

HEAD GATES: (See FEEDER GATES.)

HEADWAY: A command shouted by the captain of a canal boat when he decided that the boat had sufficient forward momentum to enter a lock chamber on its own. At the command of "headway," the bowsman would detach the towline from the deadeye. The driver would then get the team out of the way, and the boat would be allowed to glide into the lock chamber. As it did, the bowsman, and often also the steersman, would leap to the towpath and snub the boat's forward momentum.

HEELPATH: (See BERM or BERM[E] BANK.)

HEEL POST: The vertical member of the lock gate nearest to its support on the lock wall, and the axis upon which the gate pivoted. The heel post was rounded to fit into the hollow quoin in which it revolved. This was sometimes called a "quoin post."

HIGH WATER WASTEWAY: (See WASTEWAY.)

HOLD: (See MIDSHIPS.)

HOGGEE: A term originally used in Great Britain to describe a lower class of workmen who dug the canals. Later possibly used on some eastern U.S. canals as a name for a driver, there is no record of this term ever being used on the Ohio & Erie Canal.

HOLLOW QUOIN: The semicircular recess cut into the lock wall in which the heel post rotated as the lock gate was opened or closed.

HOUSE: (See CABIN.)

HULL: The lower body of a canal boat.

LATERAL CANAL: A branch canal that often paralleled the main canal, but on the opposite side of the river, to reach a bypassed town. Examples are the Lancaster Lateral Canal (later part of the Hocking Branch Canal) and the New Philadelphia Lateral Canal.

LEAD ANIMAL: The animal first in line in tandem hitching. When tow animals were tandem-hitched, the lead animal had to be trained to lean away from the canal on inside turns, taking up much of the force exerted on the towline by the boat and assisting the trailing animals so that they would not be pulled into the canal.

Canal boats loading at Lancaster on the Hocking Branch Canal. (From the author's collection.)

LESSEES: The consortium of Ohioans who leased all of the Ohio state canals from June 1862 to December 1877. Ostensibly six men, this actually was a group of more than twenty. They were offered a share of the company to keep them from forming many separate companies and forcing the price up. There was a strong opinion in later years that the canal system was allowed to deteriorate under the lease and was useless when returned. Many of the major canal structures such as locks and aqueducts were already in a deteriorated condition when the lessees obtained them. The lessees appear to have performed everyday maintenance more efficiently than did the state forces; however, most of the repairs were not of a permanent nature, and the system did require a complete refurbishment within a few years of being returned to the state.

LEVEL: The stretch of canal channel between locks. Though all canal-engineering textbooks call for perfectly level channels between locks, in practice, the Ohio & Erie Canal, as well as most other systems, was constructed with a specific slope per mile built into the levels. The Ohio & Erie Canal had a designed slope of one inch to the mile to ensure an even flow of water throughout.

Lift: The amount of change in elevation afforded by a lift lock. On the Ohio & Erie, with just a few exceptions, lifts ranged from six to twelve feet.

Lift Lock: A lock chamber in which a boat could change elevation. A lift lock isolated a boat from the upper or lower level and, by the inflow or outflow of water, caused the boat to be raised or lowered between levels.

Light Boat: A traveling canal boat that was carrying no cargo.

Line Boat: A type of craft with the cargo hold totally enclosed by a full cabin or house. This was the primary boat design during the early operating days on the canal when most cargo was carried in barrels, boxes, or bushels. It was also the type of craft employed by the formal freight lines. These craft were often fitted up with a bow section for carrying passengers while the crew slept in the stern section and the cargo was carried in the midships. This type of craft could also be called a "two-decker."

Lines (Companies): Generally a loose affiliation of forwarders, warehousemen, and boat owners who contracted to handle most of the freight and passenger traffic on the early canal. These lines appear to have died out sometime prior to the state reacquiring the canals from the lessees.

Lines (Ropes): There were three basic lines used on Ohio & Erie boats: the towline, the bowline, and the stern line.

The two-decker canal boat, *Wave*, on the Feeder Canal in Columbus in 1905. The *Wave* was a packet running between Columbus and Portsmouth in the early 1870s. (Photo by Pearl Nye, from the Canal History Collection of the University of Akron.)

LINK: A surveyor's length that was used in laying out the early canal systems. One link = 7.92 inches.

LOCK: (See LIFT LOCK and GUARD LOCK.) A stone, composite, wood, earth, or concrete chamber, closed at each end by miter gates.

LOCK CHAMBER: The volume of space between the closed gates of a canal lock.

LOCK CULVERTS: Masonry culverts built into the upper-berm side walls of a lock to fill the chamber from the upper level without water passing through the gates. The British called these "ground culverts." They were copied by New York's canal engineers and, in turn, by engineers on the Ohio & Erie. Lock culverts soon proved to be an operation problem, with no regularly assigned locktenders, and they were supplanted by regulating weirs and channels in 1827. Lock culverts made a reappearance on most locks with high mudsills during the early 1900s rebuild, but the end of canal operations occurred too soon to properly evaluate them.

LOCK GATES: (See MITER GATES and DROP GATES.) Wooden structures that could be closed to isolate a lock chamber from the rest of the canal.

LOCK NUMBERS: All lift locks on the Ohio & Erie were numbered. The general custom was to number locks in descending order, down from each summit level, beginning with the number 1. Since the Ohio & Erie Canal contained two summit levels, there were four No. 1s, four No. 2s, and so on. The addition of a lock at Massillon in 1838 resulted in the upper Massillon Lock being numbered 5A. As a rule, guard locks were not numbered. There were a few exceptions to the numbering rule on a couple of Ohio & Erie branch canals. The lock numbers on the Walhonding Canal began at No. 1 at the canal's lowest point at Roscoe and continued to No. 13, including the numbering of the two guard locks. On the branch Pennsylvania & Ohio Canal, the locks were numbered in descending order from the summit, near Ravenna, west to Akron and east to the Portage/Trumbull county line. The lock numbers then began again at the canal's eastern junction with Pennsylvania's Beaver Division Canal and were numbered from 1 up, as the canal ascended to the Portage/Trumbull county line. To make lock numbering on this canal even more confusing, it appears that guard locks were also numbered. Boatmen used names for locks, never numbers.

LOCK PIT: The physical excavation in which a lock was constructed. No specifications for Ohio & Erie lock pits have yet been located. However, specifications for Chesapeake & Ohio Canal lock pits called for them to be dug thirty-five feet across and two feet below the bottom of the canal. Locks on the Chesapeake & Ohio were fifteen feet wide, the same width as those on the Ohio & Erie.

LOCKING THROUGH: The action of getting a boat through a lock. In going up the canal to gain a higher elevation, with water in the lock chamber at the lower level and the lower miter gates open, the captain would give the command, "headway." The bowsman would unhitch the team, and the driver would get them out of the way. The boat would glide into the lock chamber and be snubbed to a stop by the crew. The miter gates and paddles behind the boat would then be closed and the paddles in the upper gates opened. Crew members then worked with lines on the snubbing posts or pike poles on the lock walls to keep the boat from striking anything as water rushed into the lock chamber through the upper gate paddles and the boat rose. When the head of water in the lock chamber and the upper canal level were equal, the paddles were closed and the gates themselves opened. The team was then reattached to the boat's deadeye and the boat pulled out of the lock and up the canal. Other combinations of boat directions and water levels were handled in a similar manner.

LOCKTENDER: The duties of a locktender were usually to fit the lock where he was stationed. However, during the later operating days of the Ohio & Erie, the few assigned locktenders did not fit locks but instead saw to it that sluices, wasteways, and lock-gate paddles along a particular stretch of canal were operating properly. On some eastern canals, this person was known as a "level walker." Evidence indicates that formal locktenders assigned by the state at each lock on the Ohio & Erie Canal only were authorized beginning in 1838 and were officially removed from all locks except feeders, during the 1860 operating season.

MAKE A LOCK: (See FIT A LOCK.)

This freighter is carrying about as much cord wood as possible in its midships holds. (Photo from the 1909 Ohio Board of Public Works Report.)

MEDICINE SPOON: An open-ended box or bag filled with dried manure or sawdust. When passed along the bottom of a leaking boat, the "medicine" would be drawn into the leak by the inflow of water. When wet, the dry material would swell and plug the leak. This was sometimes called a "sluice box."

MIDSHIPS: The covered center cargo space between the passengers' cabin and crew's cabin on an early line boat, or the two open cargo holds between the cabins on a three-cabin freighter.

MITER GATES: The hinged wooden structures at either end of a lock chamber that isolated the chamber from the upper or lower canal level. When closed, these miter gates formed an angle (miter) that pointed upstream. The action of the head of water in the upper level pressing against the mitered gates kept them closed and relatively watertight. These gates were manually opened and closed by pushing or pulling on the balance beams built into the top of each gate.

MITER SILL: Wooden timbers, ten or twelve inches square, spiked to the flooring of the lock foundation in an angled configuration so that the lower edge of the lock gates, when closed, pressed against them. Boatmen often called them "mudsills" for obvious reasons.

MUDSILL: (See MITER SILL.)

MULE: The offspring of a horse and a donkey, often considered an ideal animal for pulling canal boats.

OHIO & ERIE CANAL: The official name of the canal, for which we are defining terms, since a legislative act of 1849. Prior to that, it was known as the Ohio Canal. This same act changed the name of all or parts of three canals in the western part of Ohio to the Miami & Erie Canal. There are five canals in this country with *Erie* in their names—one in New York, one in Pennsylvania, two in Ohio, and one in Indiana.

OUTLET LOCK: (See LOCK.) A lock chamber located at the junction of a canal segment into a slackwater pool. Outlet locks generally had a specific lift built into them. Outlet locks on the Ohio & Erie let boats into the Cuyahoga River at Cleveland, the Scioto River at Portsmouth, and later the Ohio River at Portsmouth.

PACKET: A canal boat that carried passengers. A relatively small portion of the boats called packets on the Ohio & Erie Canal were of the sleek, narrow, fast design that dominated the Erie (New York) and Main Line (Pennsylvania) canals. Early packets on the Ohio & Erie were line boats carrying freight in a center hold and had a bow cabin set up for passengers. In 1837, a through line of express packets from Cleveland to Portsmouth was inaugurated that exclusively carried passengers and mail and attempted to adhere to an ambitious schedule of 80 hours for the one-way trip. The design of the craft used in this line more closely followed that of a standard line boat than the sleeker design of the eastern canal packets. Through express packet lines were discontinued after 1842, true packet service was available from Cleveland to Akron and Massillon until 1851 and from Columbus to Portsmouth until 1874.

PADDLE: The valving in a lock gate that allowed water to fill or empty a lock chamber. These paddles were operated from the top of the gate by a member of the boat crew (when no locktender was present) using a wrenchlike device on the paddle stem. The valving in ground culverts were also called "paddles." The terms *wicket* and *sluice* were also frequently used.

PADDLE STEM: A long, iron rod extending from the pivot point of a lock gate paddle to above the top of the gate. Locktenders, when one was available, or boat crew members, using a wrenchlike devise, would turn this stem to open or close the paddle.

PAINT SCHEMES: The combination of colors and decorative trim on a canal boat. Most boats were painted white with green or black trim. The steel mill located north of Five Mile Lock near Cleveland poured so much red iron oxide into the canal, however, that many north-end boatmen painted the hulls of their craft an iron color.

PERCH: An engineering term often used in figuring amounts of stonework laid. One perch = 24.75 cubic feet.

PIKE POLES: Six- to eight-foot-long wooden poles tipped with iron points. They were used by canal boat crews to move a boat within the

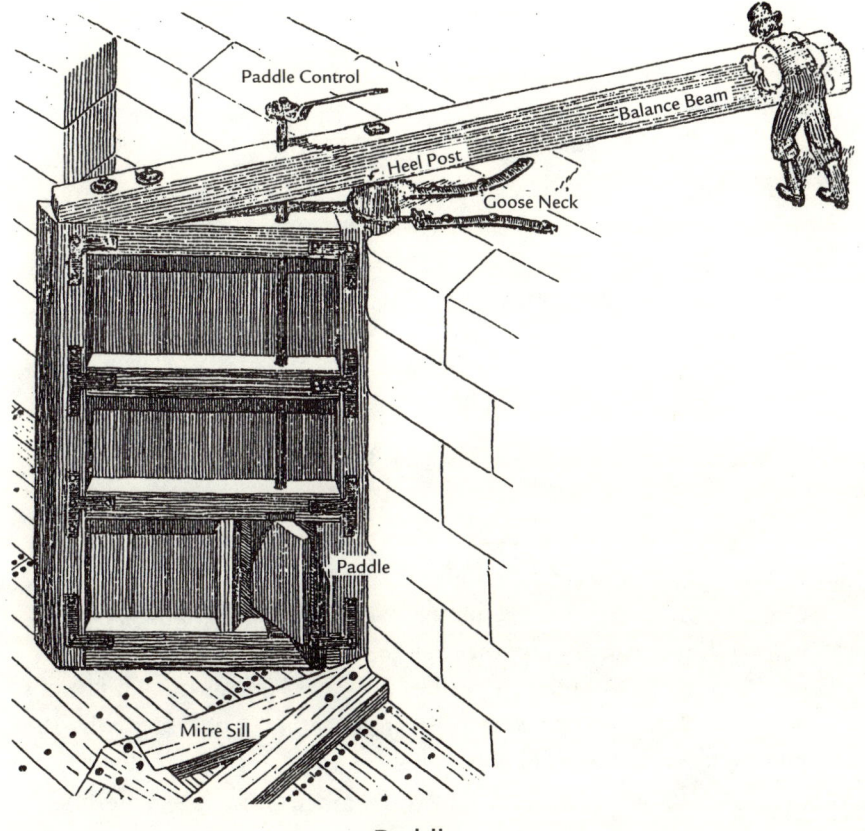

Paddle Control
Balance Beam
Heel Post
Goose Neck
Paddle
Mitre Sill

Paddle

channel by pushing along the bottom. They were also used to fend off from lock walls, docks, canal banks, etc. State boat crews abhorred them, as the iron tips could tear a channel bottom if it had been puddled, causing leaks.

PILOT: The guide of a canal boat through a tricky stretch of canal. In Akron, at the head of the staircase of locks, certain individuals would hire out to steer a boat through the difficult channels that connected the fifteen locks through town (at fifty cents a trip).

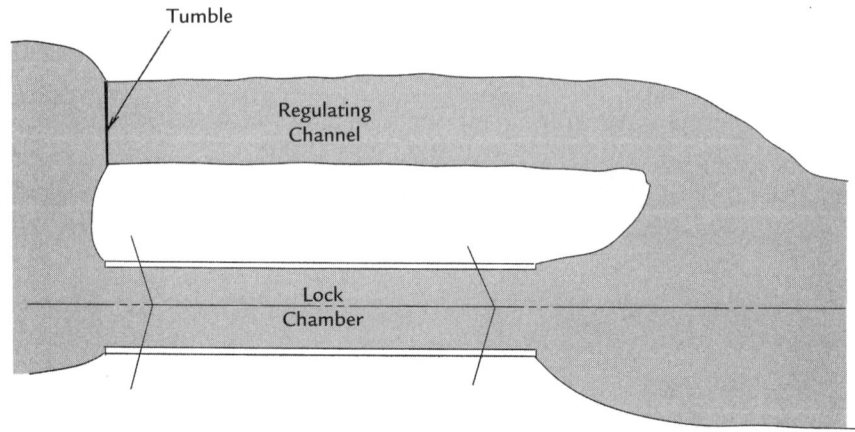

Regulating channel, plain view

PRISM: The trapezoidal, cross-sectional shape of the canal's channel. Canal engineering texts recommended a two-foot rise to a three-foot run for the interior slope of the channel's banks. New York's Erie Canal, with its forty-foot minimum width at the waterline, four-foot minimum depth, and twenty-eight-foot width at the bottom, adheres to this. The Ohio & Erie engineers cheated a bit by employing a two- to three-and-one-half-foot slope: a forty-foot minimum width at the waterline, a four-foot minimum depth, and a twenty-six-foot minimum width at the bottom.

PUDDLE: A layer (not to exceed six inches) of gravel and good clay, whetted down and thoroughly rammed with cattle's hooves or iron bars until all the air pockets had been driven out of the mixture. Several layers of puddle were used when a greater thickness was required. Puddle was often used in aqueduct, culvert, and lock foundations if there was no bedrock at the site. A thin layer of puddle could act as a watertight liner for the canal's channel through porous ground.

QUOIN POST: (See HEEL POST.)

REBUILD: There was an attempted rebuild of the Ohio & Erie Canal from Cleveland to Dresden in the early 1900s. Appropriations were begun in 1904; however, the money ran out after the 1909 season with the rebuild only completed as far south as the village of Trenton.

REGISTER: A toll keeper's logbook. A register was kept at each tollhouse on the Ohio & Erie Canal and contained the name, captain, and home port of every boat operating on the canal.

REGULATING CHANNEL: A channel around nearly every lock on the Ohio & Erie Canal that connected the upper and lower levels. This channel and the accompanying tumble were designed to ensure a continuous flow of water from one level to another and to regulate the amount of water in the levels above and below the lock. They have been variously called *sluices, spillways, tumbleways, wasteways,* and *raceways.* They were called *regulating flumes* on the Chesapeake & Ohio Canal.

REGULATING WEIRS & CHANNELS: (See REGULATING CHANNEL.)

RESERVOIR: An impoundment of water to supply the canal during dry seasons.

RIGHT-OF-WAY: The "priority rules" of the canal spelling out when a boat captain could pass or overtake another boat. In general, an upstream-bound boat had right-of-way over one bound downstream. A faster boat had right-of-way over a slower one going in the same direction. A packet had right-of-way over a freighter. The boat without right-of-way would be steered over to the heelpath, its team would move to the outside of the towpath, and the towline would be allowed to gain slack and sink to the bottom of the canal. The crew with the right-of-way would then run its team and boat over the towline between the other team and boat. Of course, all these rules were subject to interpretation and how much one crew could intimidate the other.

RIPRAP: Loose stones laid in a pattern on an earthen embankment to provide some protection against water action.

ROAD BRIDGE: A wooden bridge across the canal maintained by the canal maintenance crew. Sometimes used for dedicated roads, they more often allowed farmers access to a field cut off by the canal.

ROD: A surveyor's distance measurement, often used in laying out Ohio's Canals. One rod = 16½ feet.

ROPE FENDERS: Large-diameter (three or four inches) braided hemp ropes that were attached to the bow of a canal boat. These helped protect the planking from abrasion by the stone lock walls.

RUB RAILS: Wooden exterior ribs with iron facing located along bow and hull of a canal boat to protect the planking from rough stone walls.

SCOW: An Ohio & Erie Canal work boat that was steerable from the craft. Generally shorter than a normal boat with no cabins, it was square at the bow and stern and flat-bottomed. Some early boats on the Ohio & Erie had rounded bows to conform to design regulations, but they were slab-sided and flat-bottomed, or "scow built." These were often registered as scows.

SEASON: The length of the boating year. On the northern portion of the Ohio & Erie, it usually ran from ice breakup around mid-March or early April to the formation of heavy ice in late December.

SHEET PILING: A series of two- or three-inch-thick planks routed on the edges for tongue and groove fitting. These were driven into the earth beneath a structure's foundation and across its width to form a solid barrier, preventing water from passing underneath the founda-

Note the rub rails and towline on this canal boat as it exits Storad Lock north of Clinton. (Photo by Pearl Nye, from the Canal History Collection of the University of Akron.)

tion. The piling planks were then cut flush with the top surface of the squared-timber foundation and spiked directly to the front edge of one of the squared timbers. Two or three rows of sheet piling were run under each timbered foundation.

SHORT HAUL BOAT: A craft whose regular hauling routes were short enough that its team could be stabled at their home port. As a result, these boats didn't require stable cabins and could use the entire midships to carry cargo.

SIDECUT: A branch canal that extended from the main canal to a specific town or location. An example is the Dresden Sidecut that connected the Ohio & Erie Canal to the Muskingum River at Dresden.

SLACKWATER: An area of water impounded behind a dam. Occasionally, an entire canal system was referred to as "slackwater."

SLACKWATER CROSSING: A point where a canal channel entered and crossed a stream in the impounded water (slackwater pool) behind a dam.

SLIP: A small cutoff or sidecut whereby boats could get off the main canal to service an industry, such as a coal mine or quarry. Two examples of slips on the northern division of the Ohio & Erie Canal are the Peacock Slip, located just above Clinton, which serviced a mule-drawn tramway from the Peacock Mine in Rouges Hollow, and the Crawford Slip, about a half mile below Butter Bridge Road in Stark County, which serviced a tramway from the Crawford Mine across the Tuscarawas River.

SLUICE: An engineering term for a structure that allowed a controlled amount of water out of an impounded area (for example, a canal channel or reservoir). This term has been used to describe various structures from locks to gate valves. The term was most consistently used on the Ohio & Erie to describe the device that allowed a controlled amount of water out of a canal channel. There were one or more small sluices on each level along the canal. Three examples of big sluices on the northern division of the Ohio & Erie are above Five Mile Lock near Cleveland, at Beaver Run near Navarre, and at the Sugar Creek crossing near Canal Dover. These devices have been termed *waste gates* or *floodgates* in numerous official reports.

SLUICE BOX: (See MEDICINE SPOON.)

SLUICE WAY: (See REGULATING CHANNEL.)

SNUBBING POSTS: Two round, wooden members, partially embedded in the earth, adjacent to each lock on the towpath side of the canal.

The state boat, *Simon Perkins,* being launched in Akron in 1909. (Photo from the 1909 Ohio Board of Public Works Report.)

These were used by crew members with lines to arrest the boat's forward momentum (snub) as it entered the chamber.

SPILL: (See SLUICE or TUMBLE.)

SPILLWAY: (See REGULATING CHANNEL.)

SQUARED: The condition of a lock that couldn't be used without fitting. A lock was squared when the last boat through had been going in the same direction as the boat about to enter. Because the lock wouldn't allow passage of two boats going in the same direction without fitting, it was said to be "squared off" against the second boat.

SURFACE WASTEWAY: (See WASTEWAY.)

State boats did not need center stable cabins. (Photo from the 1909 Ohio Board of Public Works Report.)

STABLE CABIN: The center cabin on a three-cabin freighter that housed the spare team while the boat was operating and all of the animals when the boat was idle.

STAIRCASE OF LOCKS: A euphemism for many individual locks in a relatively short stretch of canal. Akron's staircase of locks consisted of fifteen in a one-mile stretch. There was also a staircase of eight at Lockville and five at Lockbourne.

STATE BOAT: A boat maintained by the state as a work craft. They were put into commission after the canal system was returned to the state by the lessees. Such craft were similar to a three-cabin freighter but did not have the center stables. Each state boat was assigned a section of canal close to its home base where the team could be stabled.

The stern deck of abandoned State Boat No. 2. (Original photo by John Wunderle of Cuyahoga Falls.)

STATE CREW: The crew of a state boat, normally consisting of a captain and a number of workmen. One (usually the newest) crew member was designated to stay on board all winter with the acting captain to see to winter maintenance.

STEERSMAN: A crew member of a canal boat who "manned the tiller." It was his duty to steer the craft and to maintain a clear distance from the towpath and berm banks. It could also be his duty to handle the stern snubbing lines when locking through.

STERN: The rear portion of a canal boat.

STONE LOCK: A lock chamber constructed entirely of stone. Initial 1825 specifications called for an interior chamber of cut, finished stone so that, with the backing stone, a lock wall would measure no less than five feet thick at the bottom and four feet thick at the top. The walls were additionally supported by buttresses (four feet square to the upper waterline) every twelve feet. By 1832, buttresses were no longer called for. The interior of these stone lock chambers was made of finished stone set in courses of headers and stretchers. The headers tied together the interior and backing walls. The spaces between all stones were filled with small pieces of stone and well-worked puddle. A renewable flooring of two-inch plank was spiked to the foundation floor planking between the stone walls.

STOP BOARDS: Planks that fit into special grooves at the head of most locks. By inserting these planks, the flow of water into the lock could be stopped and maintenance performed on the lock chamber or gates without draining the entire stretch of canal.

STRETCHER: A stone block used in construction—as in the walls of a lock chamber—whose longitudinal axis lay parallel to the length of the lock. In 1825, the minimum thickness of these stones was to be held at twelve inches with a minimum width of fourteen inches. Later specifications called for a width of thirty inches. The 1825 specifications called for no minimum length, but later specifications called for a minimum length of five feet.

SUMMIT LEVEL: The highest level on a canal where it crossed from one river valley to another. Actually, there were two summit levels on the Ohio & Erie Canal, as that canal made two headwater crossings: the Portage Summit near Akron and the Licking Summit near Newark.

SWEEP: (See BALANCE BEAM.)

SWELL: In cases of a heavily loaded or waterlogged boat, combined with a high mudsill, and with the lock full of water from the upper level plus the boat, the lower gates would quickly be opened (without lowering the water level through the wickets) and the boat allowed to rush out, over the mudsill, on the swell of water. This was to "swell" a boat.

SWELLING: A method for getting canal boats off sandbars or removing small bars from the canal channel. The first lock above the problem area was allowed to fill with water while the lower gates were kept closed. Both lower gates were then opened at once without first opening the paddles. The resulting swell of water would wash the boat off the bar or the bar out of the channel—sometimes. Also referred to as "washing."

SWING MULE: The trailing mule (or horse) in a tandem-hitched team.

TANDEM HITCHING: Hitching towing animals one behind the other. This was done because the towpath on the Ohio & Erie was, at a minimum width of ten feet, not wide enough for teams of two or three animals hitched abreast. Near the end of operation on this canal, when traffic was extremely light, a few boatmen did run with abreast-hitched teams, as more power could be generated by a given number of animals that way.

TEAM: The set of animals used to tow canal boats. Generally, two or three animals hitched in tandem were required to pull a standard three-cabin freighter and approximately eighty tons of cargo. One animal could pull a light boat. Express packets would use three or four horses.

TILLER: A movable blade, controlled from the stern deck of a canal boat, by which the steersman could angle the rudder and direct the boat.

TOE POST: The vertical member of a miter lock gate at the intersection of two gates of a set, closed, pointing upstream. The toe post was mitered, or cut to its final angle, after the gates were hung to ensure a tight fit.

The tandem-hitched mules are pulling this freighter at the end of a towline. (Photo from the Canal History Collection of the University of Akron.)

TOLL: The amount of money charged for transporting cargo and passengers along the Ohio & Erie Canal. The per-mile rate of toll was usually in the form of pennies or mills, per weight or container (which changed over the years as conditions changed), and was paid at each tollhouse along the canal.

TOLLHOUSE: The place where boatmen paid their tolls. There was a maximum of eleven tollhouses along the Ohio & Erie Canal. These were located at Cleveland, Akron, Massillon, Canal Dover, Roscoe, Newark, Carrol, Circleville, Chillicothe, Waverly, and Portsmouth.

TOLL KEEPER: The employee who accepted tolls for the state at a tollhouse.

Two canal boats showing off their newly painted transoms. (From the author's collection.)

Towline: A hard, braided, hemp line, ¾" in diameter, between 150 and 200 feet long, by which animals pulled canal boats through the canal channel.

Towpath: The path parallel to the canal, generally along the river side, along which the drivers and animals trod, pulling the canal boat on a towline. The towpath of the Ohio & Erie Canal was a minimum of two feet above the waterline, a minimum of ten feet wide, and sloped slightly away from the channel. A towpath was constructed along only one side of the Ohio & Erie Canal.

Transom: The stern of a boat upon which the boat's name or other "scenery" could be painted.

Two-decker on the southern stretch of the canal. Note rope fenders and rub rails. (From the author's collection.)

TRIPLE LOCK: (See DOUBLE LOCK.) There was one triple lock at the head of the Walhonding Canal at Roscoe.

TUMBLE: A waterfall-like structure built into each regulating channel. It allowed water from the upper level to flow into the lower level in a controlled manner while maintaining a proper head of water in the upper level. These structures were less frequently called "weirs" and "spills."

TUMBLEWAY: (See REGULATING CHANNEL.)

TWO-DECKER: (See LINE BOAT.)

VALVE: (See PADDLE.)

WASHING: (See SWELLING.)

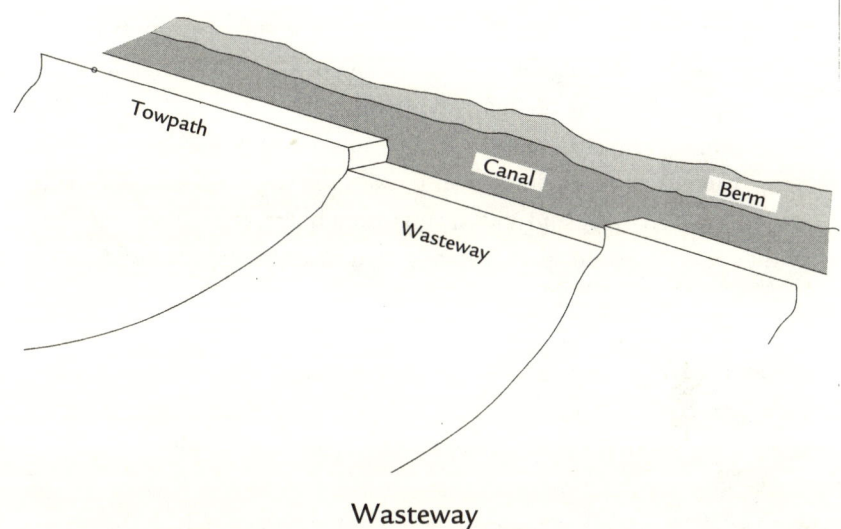

Wasteway

WASTE GATES: (See SLUICE.)

WASTEWAY: A cut-down section of the towpath lined with riprap or concrete that allowed excess water to flow out of the canal. At least one wasteway was evident on most levels on the Ohio & Erie Canal in addition to sluices. Wooden towpath bridges carried teams over these low spots after the 1900s rebuild, but there is some evidence that teams were forced to walk through these cut-down sections and get their feet wet during times of high water in the early days of the canal. This structure was also variously called a *waste weir,* a *high water waste weir,* and a *surface waste weir.* The term *wasteway* was also used in some instances to define what this glossary calls a *regulating channel.*

WASTE WEIR: (See WASTEWAY.)

WAY STATION: A formal facility along the Ohio & Erie Canal that supplied spare teams. They were maintained at intervals by the early freight and passenger lines. The freight-line way stations appear to have been phased out during the financial panics of 1837 and 1841. At least one express packet line maintained way stations along the entire length of the canal (approximately twelve to sixteen miles apart) from 1837 through 1842. There is some evidence that packet-line way stations were maintained between Cleveland and Akron through 1851, though some packets of that period carried spare teams on board, as did many freighters.

WEIGH LOCK: A lock structure with a built-in scale used to determine the loaded weight of a canal freighter. The weight of the empty boat was kept on record, and the two weights were used to determine the tolls for each trip. One of these weigh locks was located at Cleveland.

WEIR: A structure (temporary dam) containing a notch of known geometry that is inserted across a stream or channel so that all the water flows through the notch. The head or height of water flowing through the weir is measured. Flow rate can be calculated by entering the measured head into a water flow formula specific to the geometry of the weir. The term *weir* was sometimes used for any dam-like structure.

WHIPPLE TREE: (See WIFFLE TREE.)

WICKET: (See PADDLE.) The British blame the Americans for this term, yet most Americans think it came from the British. It appears in early European engineering texts, and several state reports use the terms *paddle* and *wicket* in the same paragraph to mean the same thing.

WIDEWATER: Often during initial construction it was cheaper to allow the canal channel to extend to the far hillside and not construct a berm bank at all. This resulting widewater, often as wide as 150 feet, could become a storage facility, a turning basin, and/or an ice-cutting pond.

WIFFLE TREE: The mechanism by which towing animals were hitched together and to the towline; some boatmen called this a "whipple tree."

WINCH: A mechanism employing two or more drums or sheaves on which a line or rope is wrapped, multiplying the mechanical advantage for hauling or hoisting.

WOODEN LOCK: A lock constructed entirely of wood. The lower portion of such a lock (which was always to be immersed in water) and the upper portion were built as separate structures. Thus, the upper portion could be renewed every six to eight years as required without affecting the lower portion. There were no wooden locks on the Ohio & Erie Canal, but at least one was used on the privately financed branch, the Pennsylvania & Ohio Canal.